Electronic Projects

Other Constructor's Projects Books

Electronic Game Projects
Electronic Projects in Audio
Electronic Projects in the Car
Electronic Projects in Hobbies
Electronic Projects in the Home
Electronic Projects for Home Security
Electronic Projects in Music
Electronic Projects in Photography
Electronic Projects in the Workshop
More Electronic Projects in the Home
Projects in Amateur Radio
Projects in Radio and Electronics

Electronic Test Equipment Projects

Alan C. Ainslie
Series Editor Philip Chapman

Newnes Technical Books

Newnes Technical Books
is an imprint of the Butterworth Group
which has principal offices in
London, Sydney, Toronto, Wellington, Durban and Boston

First published 1981

© Butterworth & Co. (Publishers) Ltd, 1981
Borough Green, Sevenoaks, Kent

All rights reserved. No part of this publication may be reproduced or transmitted in any form or by any means, including photocopying and recording, without the written permission of the copyright holder, application for which should be addressed to the Publishers. Such written permission must also be obtained before any part of this publication is stored in a retrieval system of any nature.

This book is sold subject to the Standard Conditions of Sale of Net Books and may not be re-sold in the UK below the net price given by the Publishers in their current price list.

British Library Cataloguing in Publication Data

Ainslie, Alan C
    Electronic test equipment projects.
    1. Electronic instruments – Design and
    construction – Amateurs' manuals
    I. Title
    621.3815'48   TK9965   80-41568

ISBN 0-408-00528-9

Typeset by Butterworths Litho Preparation Department
Printed and bound in Great Britain by
William Clowes (Beccles) Limited, Beccles and London

Preface

Test equipment projects have always been popular with amateur electronics hobbyists, as whatever project has been embarked upon, there always arises the need to verify performance or locate faults.

The test equipment projects described in this book are intended as key items to assist with the construction or development of both audio and r.f., as well as logic and control instrumentation circuitry. The equipment covers all the requirements of the home laboratory, with the obvious exception of a good multi-range meter.

The projects are all presented as basic units built on a single PCB (or *Veroboard*) and as such are functionally complete. However, for the instruments to be used for routine bench work, suitable housings or cabinets have to be provided. Each chapter contains suggestions as to the most suitable method of housing for each project. It is perhaps a good idea to build the projects in matching cases with similar layouts of knobs etc.

Various connectors are suitable for these projects, the 'professional' standard being 50Ω BNC. 75Ω coaxial connectors (TV plugs) are far cheaper and are robust enough for amateur use. TV coax then makes inexpensive connections unit to unit. (The signal generator project will need modifying by making the output resistor 82Ω for 75Ω connectors).

An interesting point is that several projects can be usefully combined into a single cabinet, sharing such expensive items as panel meters, power supplies, etc. to say nothing of the cost of the cabinet itself. Such a combination could be:

>millivoltmeter
>THD meter
>low distortion oscillator

to construct a self-contained distortion measuring set.

Many of the projects are suitable for further development or modification to extend the operating ranges or specifications, making them suitable for both amateur constructors and advanced engineers.

Contents

1. Power supply   1
2. D.C. millivoltmeter   8
3. Electronic resistance meter   14
4. R.F. signal generator   20
5. A.C. millivoltmeter   30
6. Capacitance meter   36
7. Audio oscillator   40
8. Square wave unit   45
9. Direct reading frequency meter   50
10. Function generator   56
11. TTL pulse generator   63
12. Total harmonic distortion meter   72

   Appendix   83

1

Power Supply

The projects described in this book each need a source of low voltage d.c. power in order to function correctly. In many cases this power can be provided from batteries but for the test equipment to be used continuously in the hobbyist's workshop a source of mains power is much more economical. The design covered in this chapter is for a mains powered unit with outputs of +1.2V to +20V and −1.2V to −20V, fulfilling the basic requirements for a bench power supply for general electronic experimentation. In addition the power supply could be used to power any of the individual projects. (Some projects require only a positive supply, and in these cases the negative part of the supply can be omitted – drawn in black on the circuit diagram).

Circuit

A bench power supply has to be adjustable over a fairly wide range of output voltage, in this case 1.2V to 20V, and hold the set output irrespective of changes in the current drawn by the load as well as variations of mains voltage feeding the unit, which could be as much as 15%. High performance power supplies use a stable reference voltage against which a portion of the output voltage is compared and adjusted when any difference occurs through load or line variations. The minimum requirements for this system are:

> a reference voltage source
> comparison amplifier
> power transistor to control the output voltage.

Additionally it is necessary to protect the devices against abuse, such as may arise should the output of the power supply be short circuited and therefore excessive output current flow.

Figure 1.1

Power supply

Recently the functions of reference voltage, amplifier, pass transistor and protection have been integrated into a single i.c. package, allowing high performace fixed and variable supplies to be constructed with relatively few components.

This design (Fig. 1.2) uses two of the very latest regulators from National Semiconductor designed to enable high performance variable supplies to be built with a minimum of components (in reality only two resistors are required!). This

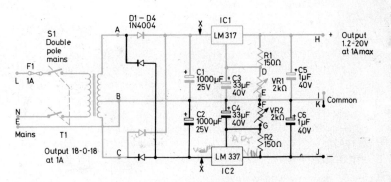

Figure 1.2

Power supply circuit diagram

allows constructors to build power supplies with a much reduced possibility of a catastrophic failure on switch on, resulting from incorrect assembly or faulty components.

The mains transformer is supplied with a.c. mains and gives a centre-tapped output of 18–0–18V. Diodes D1 to D4 form a bridge rectifier to charge C1 to around +24V and C2 to −24V. C1 and C2 remove mains frequency buzz.

The LM317 regulator (LM337 for the negative side) is a three terminal device, *in*, *out* and *control*. Internally the LM317 operates to maintain a voltage of 1.2V across R1 (R2 on the negative side). With VR1 at zero resistance the output voltage is therefore 1.2V. Setting VR1 to 150Ω gives an output of 2.4V and so on. C3 smoothes the *control* input to the i.c.

Construction

The power supply is built onto a piece of 0.1in pitch *Veroboard* 2.5in × 5in. No cuts are required in the copper tracks and the components are inserted exactly as in Fig.1.3, which shows the

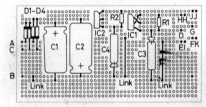

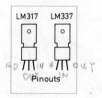

Figure 1.3

Power supply *Veroboard* layout

components' location on the board viewed from the component side. Refer to Fig. 1.1. Before commencing assembly, ensure that the copper tracks run the correct way, i.e. along the length of the board.

Insert the components one by one into their correct positions on the board, solder neatly in place and cut off excess leads. Take care to mount D1 to D4 correctly. Note that IC1 has two leads crossed over. One of these leads must be covered with a 5mm length of insulated sleeving to prevent a short circuit.

External connections are made to the board with *Veropins* at the places marked with a letter on the layout drawing.

For use as a general purpose supply for all of the projects in this book, Fig. 1.4 shows how the additional components are connected to the *Veroboard* using insulated instrument wire.

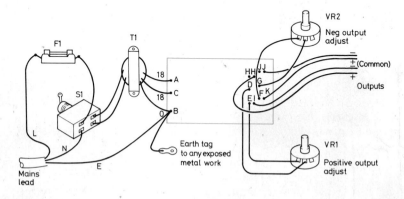

Figure 1.4

Interconnection diagram

Heatsinks for IC1 and IC2 can be combined with the requirement to house the power supply in an earthed metal case for obvious safety reasons. Fig. 1.5 shows how this is done – be sure to use insulating washers as the cooling tab on the i.c. is not isolated.

Four PCB support pillars are used to hold the *Veroboard* in the bottom of the case, while a U formed from an 8cm × 2cm strip of 16 gauge aluminium conducts heat from the two i.c.s to the side of the case. Sockets or connectors can be fitted for the outputs while two ⅜in holes will accommodate the variable controls. Where the mains lead enters the case it should be protected by the fitting of a rubber grommet and a strain relief.

Testing

With the mains connected and the unit switched on, a voltmeter connected to the positive output should indicate between 1.2V and 20V depending on the setting of VR1.

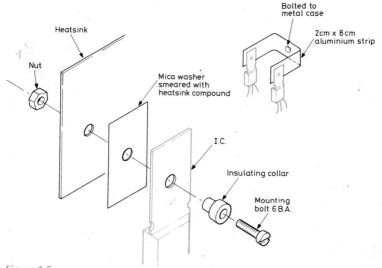

Figure 1.5
I.C. heatsinks

Conversely VR2 sets the output on the negative side. Connecting any load such as a transistor radio with the output set to the appropriate voltage should have no effect on the output voltage.

Use

This power supply will adequately power any of the projects described in this book as well as most other solid state projects.

Single positive output

As a general purpose supply, the design as described is perfectly adequate, but it is possible to effect certain economies if the supply is being used to drive only one project for which only the positive output is required. In this case, the components shown in black on the circuit diagram are simply omitted, leaving the positive section of the unit unaltered.

Where the supply is to be used only to power one particular project, it may be a useful economy to set the output voltage with VR1 and then to measure its resistance and connect in its place a fixed resistor of similar value.

Laboratory power supply

Adding output voltage meters and additional protection against abuse enables the design to be extended to a versatile laboratory power supply.

Fig. 1.6 shows what is required in addition to the basic unit. Diodes D5 and D6 protect against reverse voltage fed back into

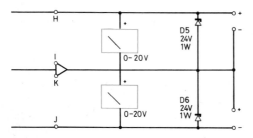

Figure 1.6
Laboratory power supply

the output terminals, perhaps from an inductive load. These components can be mounted directly across the output terminals.

Voltmeters of medium sensitivity (1kΩ/V, 1mA) are suitable and should be chosen for clear scale.

For laboratory use it is essential that the heatsink arrangements are adequate since, with the output short circuited,

Table 1.1 Components for power supply

I.C.s		Diodes	
IC1	LM317	D1	1N4004
IC2	LM337	D2	1N4004
		D3	1N4004
Capacitors		D4	1N4004
C1	1000μF, 25V		
C2	1000μF, 25V		
C3	33μF, 40V	Transformer	
C4	33μF, 40V	T1	mains voltage primary 18–0–18 at 1.5A
C5	1μF, 40V		
C6	1μF, 40V		
Resistors		Miscellaneous	
R1	150Ω	F1	1A fuse and holder
R2	150Ω	Mains lead	
(¼W unless stated)		Veroboard 0.1in pitch	
VR1, VR2, 2kΩ		Aluminium for heatsink (see text).	
		Cabinet.	

each i.c. has to dissipate something in the order of 20W. The i.c.s are protected against excessive temperature rise and will shut down should the chip temperature rise too high, but it is best not to have to rely on this 'last chance' protection.

Should it be required to add current meters, it is not a good idea to connect the meters in the leads to the output terminals as there will be a voltage drop across the ammeter which will be deducted from the actual output voltage at the terminals and load regulation will be lost.

Connecting the ammeter at the input to the regulator i.c. (point X) maintains regulation at the expense of always passing a current around 10mA through the meter, even with no load connected. However, to put this in context, it represents 1% of f.s.d. on a 1A meter and can be considered insignificant.

2

D.C. Millivoltmeter

Common multi-range meters, although quite adequate for non-critical voltage measurements, can introduce considerable errors when used to measure voltages in high impedance circuits where the resistance of the meter may be comparable to the circuit resistance. The sensitivity of a meter is expressed in kilohms per volt, and while a resistance of 1MΩ presented by a 10kΩ/V meter on the 100V range may be acceptable, a resistance of as little as 10kΩ on the 1V range would upset many circuits.

Figure 2.1

D.C. millivoltmeter

The design presented here is a sensitive d.c. electronic meter of several megohms input resistance and covering ranges of 5V f.s.d. to 5mV f.s.d. Designed to be used with a 1mA meter movement, the unit can be used as a pre-amplifier for a conventional test meter when switched to its 1mA range (or 5V range).

Circuit

A basic sensitivity of 5V f.s.d. was selected as above this even the least sensitive meters have adequate internal resistance. Ranges of 5V f.s.d., 0.5V, 50mV and 5mV f.s.d. are provided, all at the same input resistance of several MΩ.

A 741 operational amplifier is used (Fig. 2.2) in its non-inverting mode, i.e. with the input connected to the positive

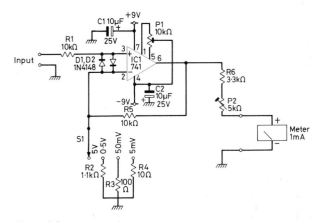

Figure 2.2

D.C. millivoltmeter circuit diagram

(+) input. R1 together with D1 and D2 protect the i.c. from damage should the input be inadvertently connected to a higher voltage. Preset P1 balances the offset so that at zero voltage input, there is zero voltage output on pin 6.

On the 5V range S1 does not contact any ground resistor and the mode of the i.c. is as a simple voltage follower with feedback via R5. 5V input will give a corresponding output of exactly 5V on pin 6 of the i.c. R6 and P2 are such that exactly 1mA will then flow through the meter causing full scale deflection of the pointer.

When set to the 0.5V range, S1 connects R2 between the negative input of the op amp and ground. Feedback in this case is derived from the potential divider R5/R2 causing IC1 to exhibit a gain of 10. Similarly, 50mV f.s.d. is achieved with 100x gain provided by the feedback divider R5(10kΩ)/R3(100Ω). The most sensitive range of 5mV f.s.d. is obtained with a gain of 1000, the feedback in this case selected by S1, as R5(10kΩ)/R4(10Ω).

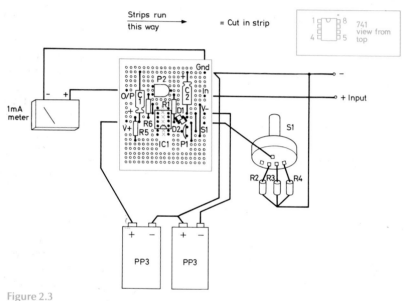

Figure 2.3

Veroboard layout

Balanced supplies of positive and negative 9V are required for this project, which is to be provided from the mains power unit or a pair of PP3 batteries.

Construction

The complete d.c. millivoltmeter is assembled on a small piece of *Veroboard* 2.2in × 2.5in. This is cut from a 5in × 2.5in standard piece so that the strips run along the short side. Cuts are to be made in the *Veroboard* as shown on the layout (total 8), using the special *Vero* cutter.

One by one the components are inserted and soldered in place – taking care over the polarity of D1 and D2 and C1 and

C2. Above all, ensure that the i.c. is the correct way round. Pin 1 will be identified by a small indentation or pip adjacent to it or a larger mark at that end of the i.c. package. Note that the i.c. pin numbers are as viewed from the *top* of the device. Generally 741's are sold in 8 pin DIL packages, but 14 pin devices can be used. A 14 pin 741 was used in the unit photographed. Pin 3 on the 14 pin package is simply located where pin 1 should be for the 8 pin package. All the connections will then be correct.

P2 is shown on the layout as a horizontal type, whereas P1 is a miniature vertical. This is because there is not adequate room for a horizontal component in this location.

Wire links are as shown on the layout (total 4). 22swg tinned copper wire is suitable – or possibly offcuts from the leads of components after soldering.

The project is completed by wiring to S1 and the meter as shown in Fig. 2.2. Two PP3 batteries are also shown so that the correct method of connecting split power supplies can be observed. Rather than unclip the batteries each time after use it is better to put a double pole switch in the positive and negative supply lines.

Calibration

Connect the batteries and switch on. Short the positive and negative input connections together and switch S1 to the 5mV f.s.d. range. Slowly trim P1 to bring the meter needle exactly to zero. This adjustment will be quite critical and may need trimming occasionally as the batteries go flat.

With S1 now set to 5V f.s.d., a 4½V battery can be connected to the input and P2 adjusted so that the meter indicates the same voltage as measured by a conventional meter connected across the same battery. Calibration is not necessary on the other ranges and is dependent on the accuracy of R5, R2, R3 and R4.

Use

The d.c. millivoltmeter will allow reliable measurement of low voltages found in transistor equipment with negligible circuit loading. The 5mV range opens up the possibility of very accurately balancing circuits as well as investigating the very low voltage outputs of may transducers, such as strain gauges, without resorting to an amplifier.

The 10:1 range switching has been chosen because the ideal 10 : $\sqrt{10}$: 1 sequence requires inconvenient resistance values. In any event, rescaling a meter face accurately to $\sqrt{10}$ (approximately 0–3) is a very tedious task. A reasonable compromise is to extend S1 and switch in a 10 : 2 : 1 sequence for which the range resistors would be:–

5V	∞	50mV	101Ω
1V	2.5kΩ	10mV	20Ω
0.5V	1.1kΩ	5mV	10Ω
100mV	204Ω		

Meters of basic sensitivity other than 1mA may be used, simply calculate the total value of R6 + ½P2 from

$$R6 + \tfrac{1}{2}P2 = \frac{5}{\text{f.s.d. (in amperes)}}$$

Table 2.1 Components for d.c. millivoltmeter

I.C.s
IC1   741 op amp 8 or 14 pin

Diodes
D1   1N4148
D2   1N4148

Resistors

R1	10kΩ	P1	10kΩ
R2	1.1kΩ 1%	P2	5kΩ
R3	100Ω 1%		
R4	10Ω 1%		
R5	10kΩ 1%		
R6	3.3kΩ		

(¼W unless stated)

Capacitors
C1   10μF, 25V
C2   10μF, 25V

Meter
1mA f.s.d. or multimeter on 1mA range.

Miscellaneous
Veroboard
S1   Single-pole 4 way rotary switch.

If the series resistance of the meter is significant, remember to deduct it from the value found for R6 and ½P2 in total. Ideally, R6 should be about twice the value of the whole resistance of P2.

Example A 50µA meter. 2kΩ resistance.

$$R6 + \tfrac{1}{2}P2 = \frac{5}{0.00005} = 100\text{k}\Omega$$

In this case R6 should be 82kΩ with a 50kΩ pot for R2.

When using the d.c. millivoltmeter as a pre-amplifier for a normal multimeter on the 5V range, R6 and P2 should be shorted out as the multimeter will incorporate its own series resistor to set f.s.d. In this case, the only calibration adjustment required will be the adjustment of P1 to maintain an exact zero on the most sensitive range.

3

Electronic Resistance Meter

Resistors are undoubtedly the most common component in electronics assemblies and there is a great need to be able to measure resistors with ease and accuracy.

Non-electronic test meters measure resistance by connecting the unknown in series with a voltmeter as shown in Fig. 3.1. As an example, a 1mA meter with a 1.5kΩ series resistor is a

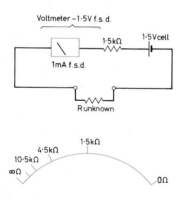

Figure 3.1

Circuit of non-electronic resistance meter

voltmeter of 1.5V f.s.d. A 1.5V cell connected directly therefore gives full deflection of the meter. Connecting the cell in series with the unknown resistor shows the first disadvantage of this simple system: the scale is reversed in that zero ohms is at the right hand end of the scale and ∞ is at the left hand end stop.

The second disadvantage is that the scale is nonlinear, giving a reading of one half f.s.d. for an unknown resistance exactly

Figure 3.2

Electronic resistance meter

equal to the voltmeter resistance – one quarter of f.s.d. is at three times the meter resistance, etc. This is a serious disadvantage for an amateur constructing a meter as repetitive calculations and intricate scale markings are required to calibrate the meter face.

The electronic resistance meter (Fig, 3.2) gives a linear scale with zero ohms at the left hand end of the scale, requiring no major rescaling of the meter face.

Circuit

IC1, a 741 operational amplifier (Fig. 3.3) forms a high impedance voltmeter of 0.5V f.s.d. as in Fig. 2.2.

Transistor TR1 is connected as a constant current source passing a fixed current through RX. The value of the current is set by D1 and the resistor selected by S1. For example, as shown the voltage between the base of TR1 and the positive supply rail is defined by D1 at around 5V. Neglecting the voltage drop across the emitter to base junction of TR1, the voltage across the resistor selected by S1 must also therefore be 5V. As shown S1 selects R2, 100Ω.

By Ohm's law, current flowing in R2 = 5/100 = 50mA.

This same current flows in TR1 collector and, therefore, through RX. It is clear that the voltage across RX is given by Ohm's law again:

V = 50mA × R, so if R = 10Ω
V = 0.05 × 10 = 0.5V, i.e. f.s.d. on the voltmeter section

The voltmeter can therefore be calibrated directly in resistance, with a linear scale ranging from 0 to 10.

By selecting different emitter resistors for TR1 with S1, the constant current source is changed, enabling resistances up to

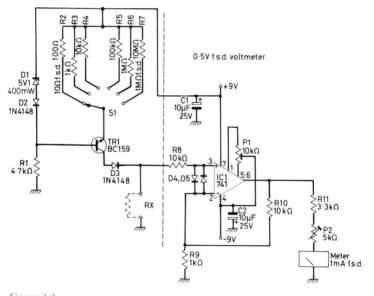

Figure 3.3

Circuit of the electronic resistance meter

1MΩ to be measured. To ensure that the constant current source is predictable and does not depend on the temperature of TR1, diode D2 is included to balance the Vbe of TR1 and render changes in Vbe insignificant.

Incidentally, by limiting the voltage dropped across the test resistor to less than 0.7V, silicon transistor junctions will not switch on. It is possible, therefore, to make measurements on resistors in circuit, without the measurements being affected by transistors, etc.

Construction

The complete electronic resistance meter is assembled on *Veroboard* 2.5in × 2.8in, with the strips running along the length of the piece (this is the remainder of a 2.5in × 5in standard piece of board left over after Project 2 has been built). Cut are made in the board at the points marked on the layout (Fig. 3.4) with a special cutter (total 12 cuts).

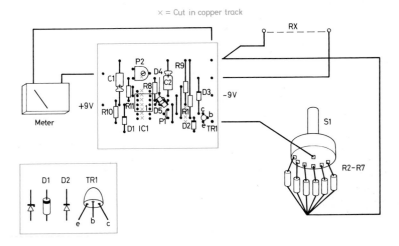

Figure 3.4

Veroboard layout

One by one the components are inserted and soldered in place, taking care not to confuse the zener diode, D1, which looks very similar to the 1N4148 diodes. Transistor TR1 must be connected correctly, carefully identifying the lead outs before inserting.

The absolute output of the constant current source depends on R2 – R7, which should all be close tolerance types (1%) or selected, to ensure good transfer from range to range. By using a 1mA meter it is quite likely that no rescaling will be necessary at all – if the meter reads 0 to 10, all that is required is to mark the switch S1 with the appropriate multiplication factor, e.g. Rx10kΩ for the 10kΩ f.s.d. range.

A double pole on/off switch will be required but is not shown on the layout for the sake of clarity. Unlike a voltmeter, the electronic resistance meter will only be required to make readings of short duration and battery life will be extended by using a simple two pole push switch, which is simply pressed to take a reading.

The test sockets or leads need a certain amount of consideration if they are not to introduce errors through their self-resistance. The problems with plugs and sockets is likely to be intermittent resistance of possibly a few ohms: results will be more reliable with a pair of permanently connected test leads fitted with substantial crocodile clips. Special flexible test lead wire is available. Thin instrument wire will be unsatisfactory as a pair of one metre leads will have a resistance of around 1Ω.

Calibration

With a dual 9V supply connected, P1 is adjusted so that the meter indication is exactly zero when a short piece of wire is substituted for RX. Removing this wire will then cause the needle to run up to the end stop at f.s.d. Select the 10kΩ f.s.d. range and connect a known accurate 10kΩ resistor as RX. Trim P2 for f.s.d. Calibration is now complete.

(An accurate 10kΩ resistor can be purchased either as 0.5% metal film for only a few pence or possibly selected from a range using a digital meter. Beware of using an average of a batch of several resistors of say 5% because in all events it is quite likely that the resistors nearest to the marked value may have been selected out at manufacture to be sold at a higher price!)

Use

Operation of the electronic resistance meter is very straightforward. The test clips are attached to the unknown resistor and the meter switched on. S1 is then adjusted to a convenient range and the value of the resistor read directly. Silicon transistor junctions will not affect readings but this clearly makes the instrument of no value for measuring transistors, other than testing for shorted junctions.

The instrument is reasonably protected should the meter inadvertently be used on equipment connected to power – obviously the readings obtained will be meaningless!

Supplement

The range ratio of 10 : 1 does not fully exploit the advantages of the accurate measuring techniques employed – a 10 : $\sqrt{10}$: 1 selection would be much better but suffers the previously discussed problem of scaling the meter face.

If the ranges are extended to allow selection in a 10 : $\sqrt{10}$ sequence, the extra resistors required will be 316Ω, 3.16kΩ, 31.6kΩ, 316kΩ and 3.16MΩ. These can be made up from, for example, 330Ω in parallel with a higher value resistor – the electronic resistance meter can be used with its calibrated prime ranges to select the intermediate resistors with the exception of 3.16MΩ.

A very useful multifunction electronic meter can be constructed by using the current source of Fig. 3.3. with the voltmeter circuit of Fig. 2.2 to give resistance and voltage ranges in the same instrument with appropriate switching.

Table 3.1 Components for electronic resistance meter

I.C.s
IC1   741 op amp 8 or 14 pin

Diodes
D1   5V1 400mW zener
D2   1N4148
D3   1N4148
D4   1N4148
D5   1N4148

Transistor
TR1   BC159

Resistors
R1   4.7kΩ
R2   100Ω 2%
R3   1kΩ 2%
R4   10kΩ 2%
R5   100kΩ 2%
R6   1MΩ 5%
R7   10MΩ 5%
R8   10kΩ
R9   1kΩ
R10   10kΩ
R11   3.3kΩ
      (¼W unless stated)

Miscellaneous
P1   10kΩ 50mW
P2   5kΩ
Meter 1mA or multirange meter set to 1mA range
S1   Single pole 6 way rotary.

4

R.F. Signal Generator

One of the most common radio constructional projects, an r.f. signal generator, poses quite serious problems for the home constructor with regard to mechanical provisions for a tuning scale, etc. as well as adequate screening. Home wound coils also add to the problems, with the result that many home-built signal generators are never actually completed. Not so this design!

Mechanical tuning of a variable capacitor has ben replaced by electronic varicap tuning with the new 1211 varicap diodes (Fig. 4.1), which permit a high total capacitance to be achieved

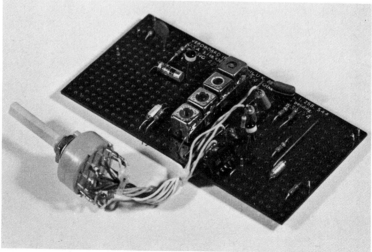

Figure 4.1

R.F. signal generator

and require only a nominal tuning voltage to attain minimum capacitance. The complete oscillator, modulator and band switching can then all be housed in a completely screened aluminium enclosure. Ready-made coils (from Toko) eliminate tiresome coil winding.

Both AM and FM are possible with this design, which as described covers 400kHz to 20MHz, although this range is easily extended.

Circuit

TR1 is biased by R1/R2 into its linear region (Fig. 4.2). The range switch S1 selects an emitter resistor R5 to R8. Each of the r.f. transformers are coupled to TR1 base and emitter by C1, and C2 to C5 as selected again by S1. The configuration can be considered as a Hartley oscillator with feedback generated by inductive coupling in the tapped primary of the r.f. transformer. To avoid loading, the r.f. output is taken from the secondary of each r.f. transformer.

Oscillation will occur at the resonant frequency of the inductance of the primary of the r.f. transformer and the capacitance of VC1 – the varicap tuning diode.

VC1 consists of two reverse biased diodes which appear as two reasonably high Q capacitors in series. The capacitance depends on the degree of reverse bias applied and is at a minimum with maximum bias. VC1 is tuned through its range by simply connecting it to the slider of a 100kΩ pot across the supply. This places a strong requirement that the supply be well stabilised to avoid frequency drift. By modulating the tuning voltage, FM can be introduced – this will be dealt with later.

A low frequency output of 470kHz is available on the 0.8 to 1.8MHz range by closing switch S2. This shorts VC1 by 470pF, dropping the resonant frequency.

An amplifier comprising TR2 is used to provide for amplitude modulation. The output inpedance of this stage is suitable to drive 70Ω co-ax cable directly or feed into a 70Ω attenuator unit.

Construction

This project is constructed on a piece of 0.15in pitch *Veroboard* (necessary to suit the pitch of the r.f. transformers) size 2½in × 5in, with the copper strips running the length of the board.

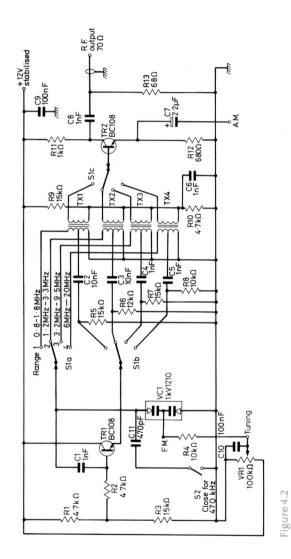

Figure 4.2

Circuit diagram of r.f. signal generator

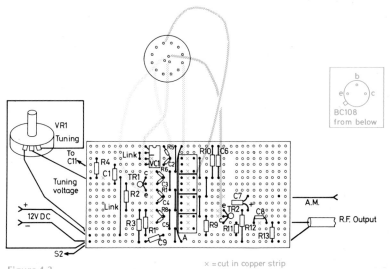

Figure 4.3

Veroboard layout

28 cuts are required in the strips as shown in Fig. 4.3. Also, 10 links are to be made with 22swg tinned copper wire.

Construction is started by carefully orientating the four r.f. transformers and inserting them onto the board, taking care to get the part numbers correct. Each transformer is in a small can itself equipped with two small earth tags. These should be removed or bent out of the way. To earth the four cans a length of tinned copper wire is routed from the extreme sides of the board adjacent to the cans (marked A in Fig. 4.3). This wire is then soldered to each can in turn (Fig. 4.4), taking care not to overheat the transformers by excessive application of the soldering iron. The rest of the board can now be made up.

The KV1210 varicap diodes are in a 5 pin package. The centre pin can be bent out of the way or cut off as it is not used in this application.

Take care to insert TR1 and TR2 correctly, and also C7. Component pairs C2/R5, C3/R6, C4/R7, C5/R8 are inserted into the board as shown in Fig. 4.4, each pair adjacent to the appropriate r.f. transformer. The junction of the two components is later wired to the switch S1b.

Range switch S1

S1 is a 3-pole 4-way switch, 15 connections in all. As a maximum the switch can be remote from the board by 6 or 7cm

To S1b

C2

R5

Component pairs C2/R5 etc are made up as shown for insertion into board

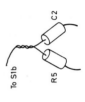

Solder

Earthing R.F. transformer cans

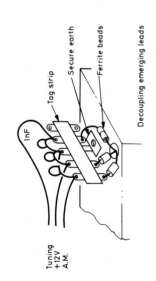

1nF

Tag strip

Secure earth

Ferrite beads

Tuning
+12V
A.M.

Decoupling emerging leads

Figure 4.4

– longer wires will introduce too much stray capacitance. Using three different colours for the wiring, the switch is methodically connected to the board. All the wires are soldered to the underside of the board except for the four connections from S1b made to the junction of C2/R5 etc. Switch S2 and C11 are added with the shortest possible leads after the board has been mounted in the diecast box.

Screening

A small diecast box, or aluminium hobby box, can be used completely to enclose the PCB and range switch. External connections to be brought into the screening box are *12V+, AM, tuning voltage* and the *r.f. output* cable. The negative supply is connected securely to the metal work of the screening box.

If available, feedthrough capacitors can be used to decouple the *12V, AM* and *tuning* VOLTAGE connections where they leave the screening box. Failing this, 1000pF decoupling capacitors and ferrite beads should be located as close to the emergence of the leads as possible (Fig. 4.4). The hole in the screened enclosure should be as small as possible to reduce r.f. leakage.

The r.f. output co-ax cable needs no decoupling but should be a tight fit in the hole in the screening box.

Calibration

A simple pointer fixed to the shaft of VR1, the *tuning* control, enables a scale to be calibrated in each of the four ranges selected by S1. Should a frequency counter be available, calibration over the whole range of 800kHz to 20MHz can be accomplished easily in a matter of a few minutes. However, if no such luxury is available, it will be necessary to resort to a short wave radio.

A 10cm length of unscreened wire on the end of the *r.f. output* co-ax cable will radiate adequate signal to be picked up by a nearby shortwave set with a similar length of wire connected to the aerial terminals.

Starting with the lowest frequency range the receiver is set to 800kHz and the r.f. signal generator tuned through the band until its output is detected by the receiver. This point is marked on the scale and then the 900kHz point found in the same way. In this manner all four ranges can be calibrated with a fair degree of accuracy.

Each of the four r.f. transformers has adjustable cores, factory set to give the coverage indicated. However, should there be inadequate overlap between ranges, the cores can be adjusted *with great care* as they are rather delicate.

Applications

Radio alignment

An r.f. signal generator is indispensable for the alignment of all kinds of radio receiver, whether home built or in need of repair. The standard radio i.f. of around 470kHz is available on range 1 by closing S2, VR1 then tuning around this frequency.

10.7MHz FM i.f. is available on range 4, while front end frequencies up to 20MHz are available on fundamentals. Harmonic suppression is quite good enough to prevent incorrect harmonics being identified but nevertheless at 100MHz FM tuners can be aligned from harmonics on range 4.

Attenuator

In this basic form the output of the generator is around 100mV, too high to be injected directly into all but the least sensitive stages of a radio receiver. Fig. 4.5 shows an attenuator which by a combination of switches and a variable control drops the output to as little as a few tens of μV on fundamentals.

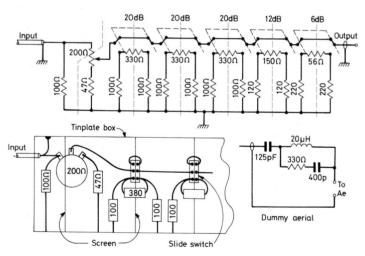

Figure 4.5

Attenuator 75Ω

The attenuator can be constructed in a small tinplate box using miniature slide or toggle switches for attenuation selection. It is essential to place a screen of tinplate between the attenuator sections to prevent leakage over each section.

To inject the signal into the i.f.s, of a radio for example, a 1000pF isolating capacitor is usually used on the end of a length of cable. However, for trimming the front end of a receiver the dummy aerial of Fig. 4.5 should be used. This loads the input to the receiver in the same way as would a normal aerial. Sets with a ferrite rod or frame aerial only require a coil of a few turns of wire placed adjacent to the aerial, connected directly to the generator output. (It is as well to remember that the aerial in the set will still be picking up broadcasts and confusing intermodulation can occur).

Amplitude modulation

To make the generator more readily identified during alignment it is normal to amplitude modulate the r.f. at a frequency of, say, 400Hz. The audio generator described in this book will directly modulate the r.f. generator – a modulation input of 100mV or so will be adequate.

Frequency modulation

By using electronic tuning it is very simple to frequency modulate the carrier. Depending on the degree of deviation required, an input of 100mV applied to the FM point via a 10kΩ resistor and 0.1μF capacitor in series will usually be adequate.

Wobbulator or frequency sweeping

If the generator is to be used for the purpose of displaying response curves on an oscilloscope by sweeping the generator over a wide range, VR1 should be replaced by the arrangement in Fig. 4.6 before the instrument is calibrated. Additional to the main tuning pot, a fine control is provided which will enable the frequency to be accurately adjusted.

The wobbulator technique uses the X output of an oscilloscope to shift the frequency of the r.f. output. Therefore each horizontal position across the c.r.o. screen represents a different frequency. As the spot moves across the screen, so the frequency changes. Applying this swept frequency to, say, the discriminator in an FM radio allows the discriminator response to be displayed by connecting the discriminator output to the

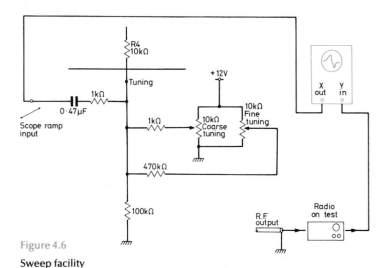

Figure 4.6

Sweep facility

y input of the c.r.o. The discriminator can then be trimmed for best response visually, rather than by tedious manual plotting.

There are only two major pitfalls to beware of when using this technique.

The sweep speed must be sufficiently slow to not distort the response given by high Q circuits

It is necessary to ensure that as the equipment is aligned and becomes more sensitive the output from the generator has

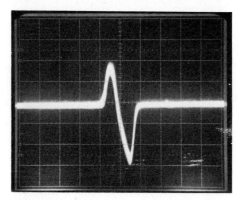

Figure 4.7

Display obtained by using signal generator swept by x-axis of oscilloscope to trace characteristics of FM discrimination at 10.7MHz

to be reduced to avoid overload and distortion of the display by the action of a.g.c. circuits in the receiver.

To reduce the need for any form of calibration, a digital frequency meter can be connected to indicate with great accuracy the trimmed output. The suppliers of the coils and varicap diode, Ambit International, market a suitable ready built frequency display module – the DFM3.

Table 4.1 Components for the signal generator

Transistors
TR1	BC108
TR2	BC108

Resistors
R1	4.7kΩ	R8	10kΩ
R2	4.7kΩ	R9	15kΩ
R3	4.7kΩ	R10	4.7kΩ
R4	10kΩ	R11	1kΩ
R5	15kΩ	R12	680Ω
R6	12kΩ	R13	68Ω
R7	15kΩ	VR1	100kΩ

(¼W unless stated)

Varicap diode
VC1	1KV1210	(Ambit International)

Coils
TX1	TOKO YMRS 8006N	(Ambit International)
TX2	TOKO KANK 3426R	(Ambit International)
TX3	TOKO KANAK 3337R	(Ambit International)
TX4	TOKO MKXNAK 3428R	(Ambit International)

Capacitors
C1	1nF	C7	2.2µF, 6.3V
C2	10nF	C8	1nF
C3	10nF	C9	100nF
C4	1nF	C10	100nF
C5	1nF	C11	470pF
C6	1nF		

Miscellaneous
S1	3P 4W rotary switch
S2	sub. min single pole toggle.

Veroboard 0.15in pitch

5
A.C. Millivoltmeter

This instrument measures the level of audio signals from 1mV to 100V f.s.d. with an input impedance approaching 1MΩ. A response of better than 15Hz–22kHz ± 0.5dB is achieved. A single PCB carries all the components with the exception of one capacitor and the actual meter movement. Although op-amps are used in the circuit, only a single 12V supply is required, making portable battery operation possible.

Figure 5.1

A.C. millivoltmeter

Circuit

The gain to give the instrument a basic sensitivity of 1mV is spread over the two stages IC1 and IC2. IC1 is a high input impedance amplifier of gain approximately 20. To allow operation from a single rail the op-amp is biased by R15 so that the output is midway between the supply rails. Resistors R2 and R3 form a resistive divider of 1000 times, attenuating the 1V, 10V and 100V inputs to 1mV, 10mV and 100mV. Further gain as well as rectification of the a.c. signal to drive the moving coil meter is provided by IC2. Sensitivity of this stage is around 20mV f.s.d. S1b attenuates the output from IC1 to cover the three basic f.s.d. sensitivities.

Construction

This project is built on a PCB, with the layout as in Fig. 5.3 (see Appendix for details on PCB construction).

Locate the components on the PCB as in Fig. 5.4 taking care over the orientation of the i.c.s and diodes as well as capacitor polarities.

The range switch is a standard two-pole six-way *Lorlin* wavechange switch which mounts directly onto the PCB. It is necessary to drill the mounting holes for the switch to larger than the usual 1mm, and 2.5mm should suffice. Take care to mount the switch securely as when the instrument is used, the switch will support the weight of the PCB and components.

If the switch is fitted with solder loops, these should be cut off before trying to mount on the PCB. Both C7 and C2 are mounted on the copper side of the board (Fig. 5.5).

Finally, insert solder pins into the board to accept the battery and meter connections, etc.

The instrument can be mounted in a convenient case using only a ⅜in diameter hole for the shaft of the range switch.

Calibration

All the ranges are tied together with only a single calibration adjustment. It is therefore possible to calibrate to virtually any standard that may be available. A known accurate meter could be used to measure the output of a 9V mains transformer, for example, and the 10V range calibrated to this value. It is

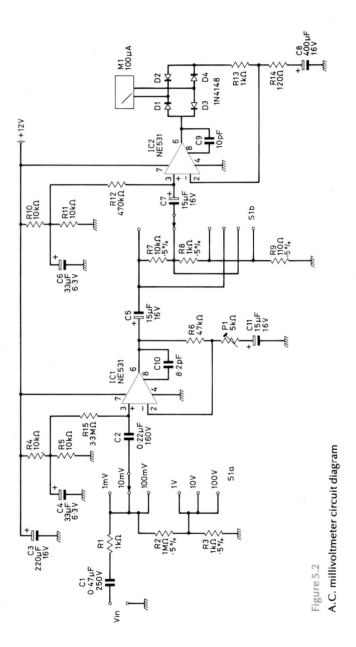

Figure 5.2
A.C. millivoltmeter circuit diagram

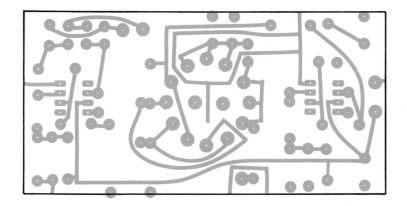

Figure 5.3

PCB layout

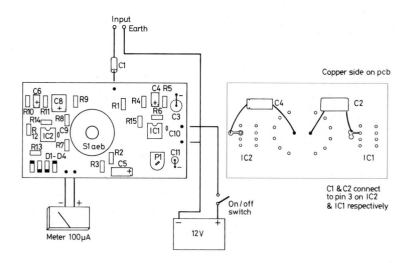

Figure 5.4

Component layout on PCB, left component side, right showing mounting of C2, C7 on copper side

important to ensure that the calibration frequency is within the range 30Hz – 10kHz for highest accuracy and, of course, it must be within the range of the reference meter.

Protection

The input stages as described are not protected to any great extent against overload. Adding two 1N4148 diodes in parallel, opposite ends together, between pins 3 and 2 of IC1 will afford adequate protection for most likely overload situations.

Range switching

Range ratios of 10:1 are not ideal, as discussed previously. If it is possible to rescale the meter movement, the simplest way to provide a $\sqrt{10}$ range is to switch a preset in parallel with R1 to increase the sensitivity of IC2 stage by a factor of 3.

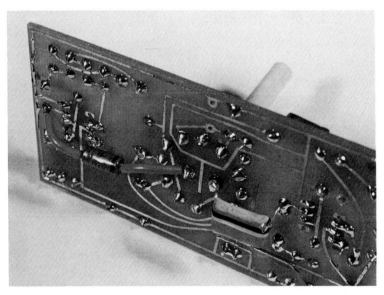

Figure 5.5
Underside of board showing mounting of C2 and C7

Table 5.1 Components for A.C. millivoltmeter

I.C.s
IC1	NE531
IC2	NE531

Resistors

R1	1kΩ		R9	110Ω ½%
R2	1MΩ ½%		R10	10kΩ
R3	1kΩ ½%		R11	10kΩ
R4	10kΩ		R12	470kΩ
R5	10kΩ		R13	1kΩ
R6	47kΩ		R14	120Ω
R7	10kΩ ½%		R15	3.3MΩ
R8	1kΩ ½%			

(¼W unless stated)

Capacitors

C1	0.47 250V		C6	33/6.3
C2	0.22 160V		C7	15/16
C3	220/16		C8	400/16
C4	33/6.3		C9	10pF
C5	15/16		C10	8.2pF

Presets
P1   5kΩ 50mW

Diodes
D1	1N4148
D2	1N4148
D3	1N4148
D3	1N4148

Miscellaneous
Meter 1mA
S1   2P 6W *Lorlin*.

6
Capacitance Meter

Capacitors are used almost as extensively as resistors in many electronic circuits, yet there are very few capacitance meters available to check these components. This design (Fig. 6.1) will simply indicate on a linear meter scale the value of capacitors from 100pF to 100µF f.s.d.

Figure 6.1
Capacitance meter

Circuit

In Fig. 6.2, IC1, a 555 timer, is connected in its astable mode, oscillating at a frequency of around 30kHz for ranges 1, 2 and 3, and a frequency of 30Hz for ranges 4, 5 and 6. R1 is relatively small and ensures that the duty cycle of the oscillation is such that the waveform on pin 3 is a series of short pulses down to 0V.

IC2, connected as a monostable, is triggered by each negative going pulse, sending pin 3 on IC2 high for a period determined by Ct and the selection of R3, R4 or R5 made by S1b. The duty cycle is therefore an indication as to the capacitance of Ct.

The waveform on pin 3 is clipped to 4.7V by D1, and meter M1 then gives an accurate indication of duty cycle and the value of Ct.

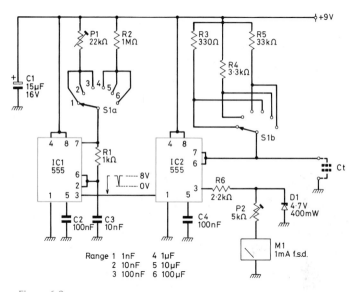

Figure 6.2

Capacitance meter circuit diagram

Construction

The PCB layout of Fig. 6.3 is used. Components are inserted as in Fig. 6.4, taking care to insert the i.c.s correctly, and C1 and D1. As in the previous project, the switch is mounted direct on the PCB and can be used to mount the instrument on the fascia panel.

Calibration

1% capacitors of various values are fairly readily available, ideally 100nF and 1µF will be required.

Initially set S1 to range 4, 1µF f.s.d. With a standard 1µF capacitor connected to the Ct terminals, trim P2 for f.s.d. on

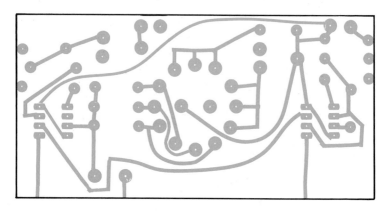

Figure 6.3

PCB layout

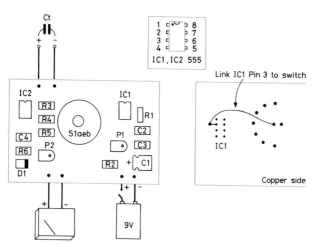

Figure 6.4

Component layout on PCB (left) and link on copper side (right)

the meter. (If the needle is a little unsteady, connect a 10μF at 10V capacitor across the meter terminals).

All that remains is to trim the first three ranges. With the instrument set to range 3, P1 is adjusted for f.s.d. with a standard 100nF connected to the Ct terminals.

A 'times 2' facility can be provided by switching a 10kΩ preset from pin 5 to ground, doubling the free running frequency of IC1. The preset is trimmed to exactly double a reading in the lower half of the scale.

Table 6.1 Components for capacitance meter

I.C.s
IC1	555
IC2	555

Resistors
R1	1kΩ
R2	1MΩ
R3	330Ω 2%
R4	3.3kΩ 2%
R5	33kΩ 2%
R6	2.2kΩ

(¼W unless stated)

Presets
P1	22kΩ 50mW
P2	5kΩ 50mW

Capacitors
C1	15μF, 16V
C2	100nF
C3	10nF
C4	100nF

Diodes
D1	4.7V 400mW zener

Miscellaneous
Meter	1mA or multirange meter on 1mA range
S1	2P 6W *Lorlin*
	PCB

7

Audio Oscillator

This chapter shows how to build a low distortion laboratory oscillator (Fig. 7.1) covering 10Hz to 100kHz with stabilised output of up to 10Vp–p. A low distortion output of below 0.008% at 1kHz is possible with t.h.d. at 20kHz only 0.03%.

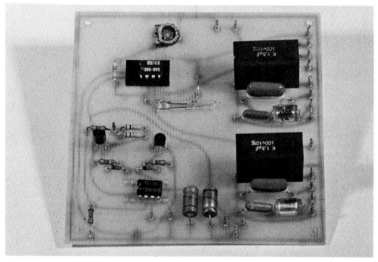

Figure 7.1

Audio oscillator

Circuit

Of the many forms of audio oscillator, the Wien bridge has been chosen, as the frequency determining components, two capacitors and a pair of variable resistors, are relatively easily

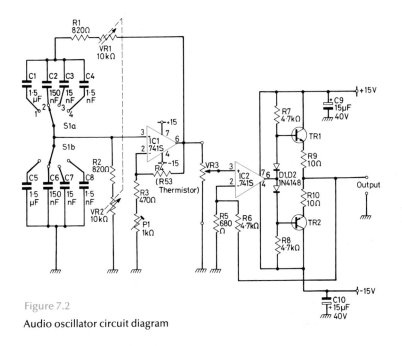

Figure 7.2

Audio oscillator circuit diagram

arranged (Fig. 7.2). (c.f. a twin tee requires a three ganged pot and switching for more capacitors).

IC1 provides the gain required for oscillation, which ideally is as small as ×3. As oscillations build up the voltage across R4 increases causing its resistance to decrease, increasing the overall negative feedback and causing the amplitude to stabilise. The output from IC1 pin 6 is controlled by VR3 and fed to IC2, an amplifier of gain approximately 8. Transistors TR1 and TR2 are connected as emitter followers to give a very low impedance output to the instrument. Feedback is taken from the output of these transistors to the negative input of IC2, reducing all distortion in the output stage to a low level. D1 and D2 provide bias for the output transistors.

Construction

The PCB in Fig. 7.3 is used to build the instrument.

Components are inserted as in Fig. 7.4 taking care to fit IC1 and IC2 correctly. The PCB layout will take both 8 pin and 14 pin devices: locate pin 1 of an 8 pin package in the position of pin 3 on a 14 pin package. TR1, TR2, D1, D2, C9 and C10 must also be installed correctly. R4, a thermistor type R53, is enclosed in a small delicate glass tube, be especially careful to hold the leads adjacent to the glass before bending into shape.

The frequency control pot, VR1/VR2, benefits from being installed with screened cable to reduce pickup of stray signals. Fig. 7.4 shows how this cable is arranged. The screen of the cable connecting pins A and B to VR2 is used to earth the

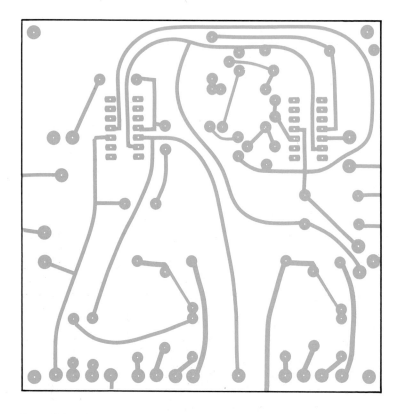

Figure 7.3

PCB layout

screens of the twin screened cable used to connect C and D to VR1. Similarly, VR3 benefits from being connected using screened cable – screen needs grounding only at one end.

Wiring S1a/S1b is quite straightforward – the wires should be kept as short as possible, however, to minimise stray pickup.

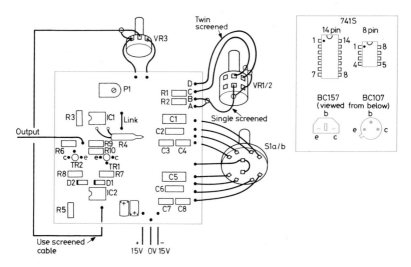

Figure 7.4

Component layout on PCB

Calibration

VR1/VR2 can be fitted with a pointer covering a 270° arc scale. All four ranges are ×10 multiples of each other, requiring only one scale to be marked. Frequency calibration is simple if a frequency meter is available, otherwise an oscilloscope can be used with the 50Hz mains as a reference. Later in the book a

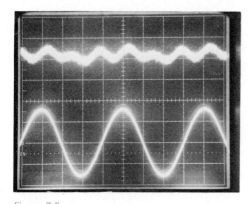

Figure 7.5

Performance of the oscillator at 1kHz. THD scale 0.003% per division. Actual THD 0.004%

direct reading frequency meter is detailed, which greatly simplifies calibration and could be incorporated in the final instrument to take the place of the frequency scale.

Preset resistor P1 is normlly used to set the output on IC1 pin 6 to 1V r.m.s., but if a distortion factor meter is available, P1 can be trimmed for optimum T.H.D.

Fig 7.5 shows a trace of the output of the oscillator when running at 1kHz.

Table 7.1 Components for audio oscillator

I.C.s
IC1   741S 8 or 14 pin
IC2   741S 8 or 14 pin

Resistors
R1   820Ω   R7   4.7kΩ
R2   820Ω   R8   4.7kΩ
R3   470Ω   R9   10Ω
R4   R53 thermistor R10   10Ω
R5   680Ω   VR1   10kΩ double gang linear
R6   4.7kΩ   VR2   10kΩ
(¼W unless stated)

Capacitors
C1   1.5μF   C6   150nF
C2   150nF   C7   15nF
C3   15nF   C8   1.5nF
C4   1.5nF   C9   15μF, 40V
C5   1.5μF   C10   15μF, 40V

Presets
P1   1kΩ, 50mW

Diodes
D1   1N4148
D2   1N4148

Transistors
TR1   BC107
TR2   BC157

Miscellaneous
S1   Two-pole, four-way *Lorlin*

8
Square Wave Unit

Whilst a sine wave signal is very useful for checking gain and fidelity of audio equipment, of course the test signal bears no relation to the signals found in music reproduction, which are by nature rarely single frequency and often transient in form. A very good indication of dynamic performance is given by square wave tests, which are a combination of frequencies, including fast transients. The unit shown in Fig. 8.1 derives a square wave from the sine wave output of the audio oscillator in the previous chapter.

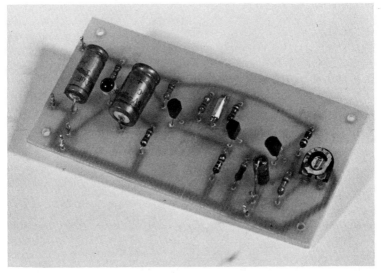

Figure 8.1

Square wave unit

Circuit

The square wave unit is based around TR2 and TR3 (Fig. 8.2), connected as a Schmitt trigger. The two transistors are connected in a regenerative loop so that when TR3 starts to turn on, it is regeneratively forced on quicker, and similarly as it switches off. This results in fast transients irrespective of how slowly the input wave form is changing. C2 speeds up the transition even further.

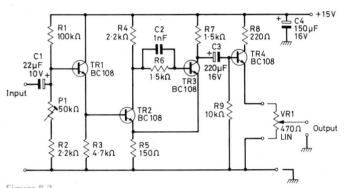

Figure 8.2

Circuit diagram of square wave unit

TR4 is connected as a low impedance follower driving the 470Ω output level control. The low impedance output ensures that the transition edges are not unduly slowed down by stray capacitance on the output, say from the use of screened cable.

A simple input amplifier comprising TR1 is used to shift the level of the input signal to the optimum point for the Schmitt trigger. P1 sets this trigger point for a symmetrical square wave.

Construction

All components with the exception of the output level control, VR1, are located on a single PCB as in Fig. 8.3. Take care to connect the transistors correctly and the three capacitors, C1, C3 and C4 (Fig. 8.4).

Power connectons are simply ground and +15V, no negative supply is required. Connections to VR1 can be short lengths of instrument wire, but by taking the precaution of using screened cable for all connections to VR1 and also the input to the square wave unit, radiation of the fast edges of square waves will be minimised. It is then possible to house the

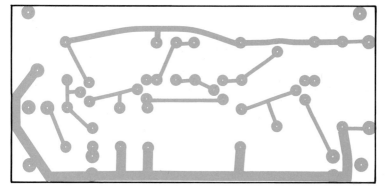

Figure 8.3

PCB layout

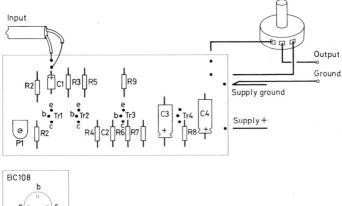

Figure 8.4

Component layout on PCB

square wave unit in the same cabinet as the audio oscillator and use both outputs simultaneously without the sine wave output picking up spikes. Earth loops should be avoided as otherwise the two outputs will affect each other.

Use

A symmetrical square wave contains even harmonics of the repetiton rate up to a relatively high frequency. Displayed on an oscilloscope, (Fig. 8.5), the risetime is so short (this instrument a few tens of nanoseconds) that it is not clearly seen. However, when passed through an audio amplifier of finite bandwidth, the risetime is increased and becomes visible.

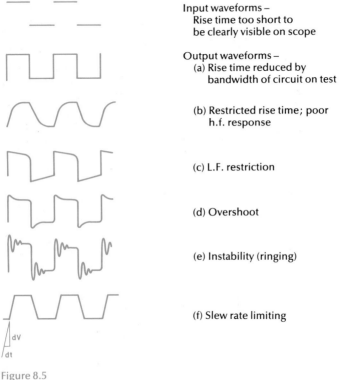

Input waveforms –
　Rise time too short to
　be clearly visible on scope

Output waveforms –
　(a) Rise time reduced by
　　bandwidth of circuit on test

　(b) Restricted rise time; poor
　　h.f. response

　(c) L.F. restriction

　(d) Overshoot

　(e) Instability (ringing)

　(f) Slew rate limiting

Figure 8.5
Square wave tests

By expanding the trace and measuring the risetime, the bandwidth can be calculated:

$$\text{bandwidth (kHz)} \simeq \frac{350}{\text{risetime } (\mu s)}$$

As h.f. restrictions become more apparent, the risetime is considerably increased. An audio amplifier tested with a 1kHz square wave should show little degradation of the leading edge.

Low frequency inadequacies are shown by a sloping top on the square wave. Again at 1kHz no slope or tilt should be discernible with an audio band amplifier extending to 20Hz.

Transient problems such as overshoot or instability are immediately obvious: audio power amplifiers often show overshoot or mild instability when driving reactive loads, such as $2\mu F$ across 8Ω.

Frequency compensation of amplifiers in an endeavour to reduce instability problems can give rise to slew rate limiting, when the output of the amplifier will only change by a certain amount per microsecond. This shows on square wave tests as a clean ramp on leading and falling edges. Each edge does not necessarily slew limit at the same dV/dt. Slew limiting is often the restricting factor in the full power bandwidth of a design.

Square wave testing is a valuable analysis tool. Apart from the tests shown in Fig. 8.5, square wave testing clearly shows up power supply deficiencies, tone control circuit deficiencies, as well as providing a clear visual clue as to overall fidelity.

Table 8.1 Components for square wave unit

Resistors
R1	100kΩ	R6	1.5kΩ
R2	2.2kΩ	R7	1.5kΩ
R3	4.7kΩ	R8	220Ω
R4	2.2kΩ	R9	10kΩ
R5	150Ω	VR1	500Ω linear

(¼W unless stated)

Presets
P1   50kΩ, 50mW

Capacitors
C1	$22\mu F$, 10V
C2	1nF
C3	$220\mu F$, 16V
C4	$150\mu F$, 16V

Transistors
TR1	BC108
TR2	BC108
TR3	BC108
TR4	BC108

Miscellaneous
PCB

9

Direct Reading Frequency Meter

Here is a simple design (Fig. 9.1) for a unit that will enable a 0–1mA meter to display frequency directly on a linear scale over the useful audio range of 20Hz to 20kHz. Measurements can be made on any waveform, with a sensitivity of less than 100mV. Applications apart from frequency measurement include tachometers and speedometers for cars.

Circuit

The input signal is amplified by IC1 (Fig. 9.2) to a level suitable for driving the Schmitt trigger IC2. Diodes D1 and D2 reduce

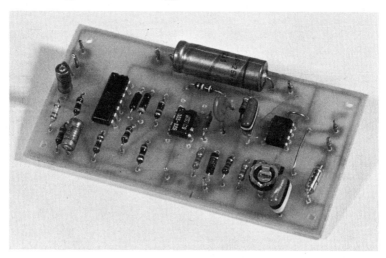

Figure 9.1

Direct reading frequency meter

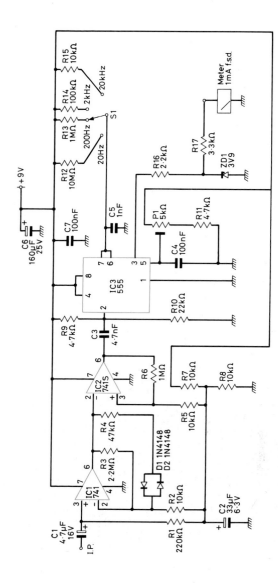

Figure 9.2

Circuit diagram of the frequency meter

the gain of IC1 for large input signals. Regardless of the input waveform, the output from IC2 is a series of pulses of the same frequency as the input.

A monostable is built around IC3, a 555 timer, triggered from the output of IC2. The output of the 555 on pin 3 is normally 0 volts, but when triggered by a pulse on pin 2, pin 3 goes high for a period dependent on the value of C5 and the resistor selected by S1. As the input frequency increases so does the proportion of time that pin 3 is high, increasing the deflection on the meter. Zener diode ZD1 ensures that the pulses passed to the meter are all of a defined level, the averaging of the meter movement then produces a deflection linearly proportional to the input frequency.

Timing of the monostable period is achieved by varying the voltage on pin 5 of the 555 with P1. This changes the overall period that pin 3 is high on all ranges and is used for calibration.

Construction

All of the components with the exception of the meter, S1 and R12/R15 are mounted on a printed circuit coard as in Fig. 9.3.

In Fig. 9.4, provision is made for using either eight or fourteen pin packages for IC1 and IC2. The pin numbers on the circuit are for eight pin devices which are mounted on the PCB in such a manner that pin 1 of the package fits into the hole

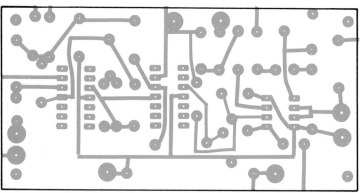

Figure 9.3

PCB layout

drilled for pin 3 of the 14 pin package. Take care to locate the i.c.s correctly. Both the diodes and the electrolytic capacitors also have to be located correctly.

Solder pins are inserted into the board for the external connections, which can be made with instrument wire, although if the input leads are long it would be better to use screened cable.

R12, R13, R14 and R15 are connected directly to the rear of S1 and in many cases the common connection can be supported on a spare tag on the switch.

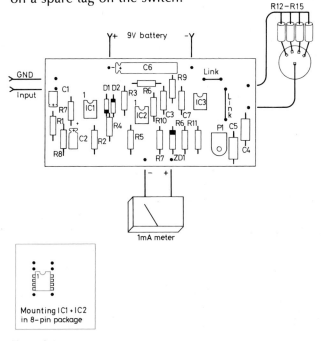

Figure 9.4

Component layout on the PCB

Virtually any convenient 1mA meter can be used or alternatively, like many of the projects in this book, the frequency meter can be assembled as an add-on unit for a multirange meter set to the 1mA range.

Calibration

Only one calibration adjustment is needed to calibrate all of the ranges. This adjustment can be made at any frequency for which a convenient standard is available, P1 is trimmed so that the meter correctly indicates the input frequency.

50Hz mains derived from a low voltage isolating transformer is a standard available to all constructors, but is not really suitable as the 50Hz point is only one quarter f.s.d. on the 200Hz range. A much better standard is the time base of a 625 line television receiver, 15.625kHz. An isolated pickup coil of say 15 turns will pick up sufficient signal when placed in the vicinity of a television line output stage.

> N.B. BEWARE OF LIVE VOLTAGES AND POSSIBLE LIVE CHASSIS.

Alternative calibration can be achieved by reference to a calibrated signal source, or an oscillator at 200Hz referenced to 50Hz mains by Lissajous frequencies on a CRO.

Use

Any signal above 100mV will be sufficient to trigger the meter which will then indicate the frequency of the applied signal. It is possible to use the instrument to indicate revolutions per minute by arranging contacts which close once per revolution (rotating magnet past a reed switch?) An extension of this would be to feed the meter from a phototransistor which would enable r.p.m. readings to be taken without contact if a contrasting line is painted on the rotating object. To eliminate the conversion from hertz to r.p.m. (×60) it is possible to recalibrate directly in r.p.m., convenient ranges achieved without changing any components other than resetting P1 being:

$$0 - 500 \text{ r.p.m.}$$
$$0 - 5000 \text{ r.p.m.}$$
$$0 - 50,000 \text{ r.p.m.}$$
$$0 - 500,000 \text{ r.p.m.}$$

Changing C5 to 10nF will change the ranges to

$$0 - 50 \text{ r.p.m.}$$
$$0 - 500 \text{ r.p.m.}$$
$$0 - 5000 \text{ r.p.m.}$$
$$0 - 50,000 \text{ r.p.m.}$$

It may be necessary to connect a capcitor of value $5\mu F$ across the meter to stop needle flutter on the lower ranges.

Table 9.1 Components for direct reading frequency meter

I.C.s
IC1   741, 8 or 14 pin
IC2   741S, 8 or 14 pin
IC3   555

Resistors

R1	220kΩ	R10	22kΩ		
R2	10kΩ	R11	4.7kΩ		
R3	2.2MΩ	R12	10MΩ	5%	
R4	47kΩ	R13	1MΩ	2%	
R5	10kΩ	R14	100kΩ	2%	
R6	1MΩ	R15	10kΩ	2%	
R7	10kΩ	R16	2.2kΩ		
R8	10kΩ	R17	3.3kΩ		
R9	4.7kΩ				

(¼W unless stated)

Capacitors
C1   4.7μF, 16V
C2   33μF, 6.3V
C3   4.7nF
C4   100nF
C5   1nF
C6   160μF, 25V
C7   100nF

Diodes
D1   1N4148
D2   1N4148

Miscellaneous
P1   5kΩ, 50mW
S1   Four-way, two-pole *Lorlin*
Meter   1mA or multirange meter on 1mA range
PCB

10
Function Generator

This instrument is a valuable source of sine, square and triangle waves, simultaneously if required (Fig. 10.1). A low impedance output of 10V p-p is available with positive and negative offsets. A frequency range of 0.1Hz to 100kHz is covered, with the facility of frequency sweeping a range in excess of 50 : 1.

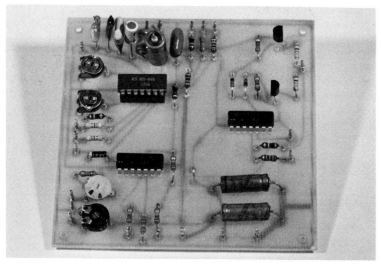

Figure 10.1

Function generator

Circuit

The design (Fig. 10.2) is based on the Intersil 8038 Function Generator chip which incorporates facilities for generating square, triangle and sine waves over a wide frequency range.

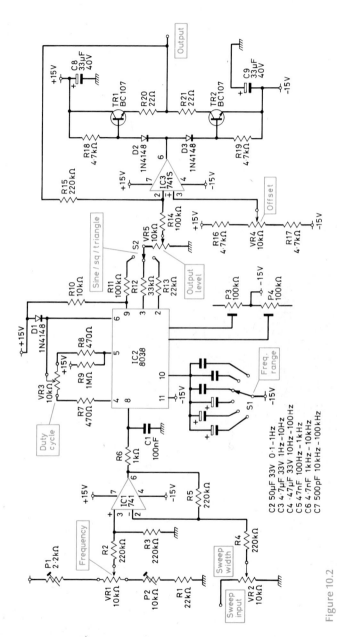

Figure 10.2
Circuit diagram

The 8038 operates by charging and discharging the capacitor connected to pin 10 in order to produce a triangle waveform. This is shaped internally to form a sine wave, whilst the charge/discharge switch is used to give a square wave output. These outputs are available on pins 3, 2 and 9 respectively.

Control of the charge/discharge currents and hence frequency is accomplished by varying the voltage on pin 8 from around +5V for high frequencies to +14.5V for low frequencies. IC1 mixes the voltage on VR1 which is of the range 5V to 14.5V with an external voltage applied to VR2, enabling external frequency sweeping over a wide range.

The actual charge and discharge currents are separated at pins 4 and 5. Varying the current on one relative to the other with VR3 enables the duty cycle to be varied so that the square wave output becomes pulse, while the triangle becomes either positive or negative running ramps Fig. 10.3.

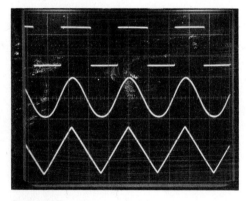

Figure 10.3

Function generator outputs – square, sine and triangle.
Adjustment of the duty cycle will additionally give + and −
pulse and + and − ramps

Diode D1 is included in the supply to the 8038 in order to ensure good performace at low frequencies when pin 8 is almost at the full supply voltage.

While the three waveform outputs are available simultaneously from the 8038, an output amplifier is provided so that the waveform selected by S2 can be amplified to 10Vp-p at a low impedance output, with facilities for shifting the level of the output waveform to operate either fully negative or fully positive, on any intermediate point.

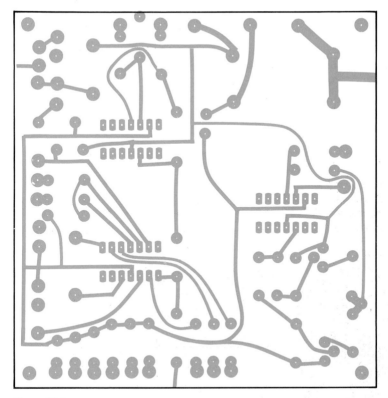

Figure 10.4

PCB layout

Construction

Fig. 10.4 shows the PCB and assembly of the complete circuit is quite straightforward as shown in Fig. 10.5. IC1 and IC3 are shown as 14 pin packages but 8 pin packages can be accommodated quite simply by inserting pin 1 of the package into pin 3 of the PCB drilling.

Take care to insert the following devices correctly:

>TR1, TR2
>IC1, IC2, IC3
>C2, C3, C4, C8, C9
>D1, D2, D3.

Solder pins are inserted into the PCB for the external connections which are made with instrument wire directly to the controls. Connections to S2 and VR5 would benefit from being screened.

Calibration

After carefully checking the board, power may be applied. An oscilloscope on pin 9 of the 8038 will confirm that the generator is running.

Setting frequency range

Set S1 to the 100Hz – 1kHz range.
Set VR3 for a symmetrical square wave on pin 9.
Set VR1 fully anticlockwise.
Adjust P2 so that the output frquency on pin 9 is 1kHz.
Set VR1 fully anticlockwise.
Adjust P1 for 100Hz output.
Repeat until no improvement is possible.

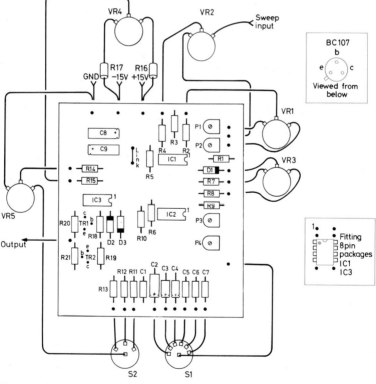

Figure 10.5

Component layout on PCB

Table 10.1 Components for function generator

I.C.s
IC1   741 8 or 14 pin
IC2   8038 Intersil
IC3   741S 8 or 14 pin

Resistors
R1	22kΩ	R12	33kΩ
R2	220kΩ	R13	22kΩ
R3	220kΩ	R14	100kΩ
R4	220kΩ	R15	220kΩ
R5	220kΩ	R16	4.7kΩ
R6	1kΩ	R17	4.7kΩ
R7	470Ω	R18	4.7kΩ
R8	470Ω	R19	4.7kΩ
R9	1MΩ	R20	22Ω
R10	10kΩ	R21	22Ω
R11	100kΩ		

(¼W unless stated)

Capacitors
C1   100nF
C2   50μF, 33V
C3   4.7μF, 33V
C4   0.47μF, 33V
C5   47nF
C6   4.7nF
C7   0.47nF
C8   33μF, 40V
C9   33μF, 40V

Transistors
TR1   BC107
TR2   BC107

Diodes
D1   1N4148
D2   1N4148
D3   1N4148
D4   1N4148

Miscellaneous
S1	One-pole, six-way *Lorlin*		
P1	2.2kΩ 50mW	VR1	10kΩ
P2	10kΩ 50mW	VR2	10kΩ
P3	100kΩ 50mW	VR3	10kΩ
P4	100kΩ 50mW	VR4	10kΩ
PCB		VR5	10kΩ

Setting waveform

Connect the scope to pin 2.
Trim P3 and pin 4 for visually best sine wave.
Check that at the lowest frequency the symmetry is maintained, if not R9 can be changed to restore symmetry by compensating for current differences on pins 4 and 5.

Use

Apart from the obvious use of the square and sine outputs, the triangle output is a very useful, convenient tool for developing all manner of linear circuitry using op-amps etc. Nonlinearity is easily detected while any desired nonlinear operation can be easily verified. Clipping and saturation are also clearly visible on CRO. At very low frequencies the operation of a circuit can virtually be followed with a voltmeter.

Frequency sweep

Rather than laboriously plotting frequency responses of filters etc. by hand it is a simple matter to display the response on a CRO by linking the X deflection to the sweep input. For a positive going X ramp the display will be high frequencies at the left, low frequencies towards the right. Selecting a suitably low sweep rate will show the response as a solid outline. (Too fast a sweep will give rise to problems when sweeping a high Q circuit, leading to erronous placement of the turnover points and possibly not displaying the full value of a peak or trough).

11
TTL Pulse Generator

When developing logic and pulse circuitry, a predictable source of pulse data is invaluable. The design illustrated in Fig. 11.1 will generate pulses up to a pulse repetition frequency of 1MHz with pulse widths down to 100ns. Facilities are provided to generate gated pulse trains, pre-pulse and delayed pulse, as well as continuous trains. Simple logic will extend the generator to provide double pulses.

Figure 11.1

TTL pulse generator

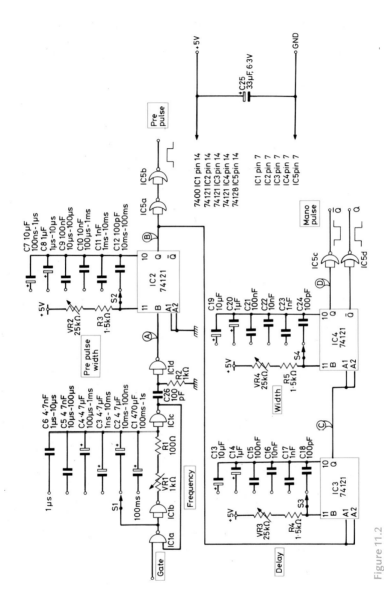

Figure 11.2
Pulse generator circuit diagram

Circuit

Three gates of IC1 are connected (Fig. 11.2) to form a simple gated astable oscillator, whose frequency is determined by VR1 and the capacitor selected by S1 over a range of 1Hz to 1MHz. The gate input inhibits the oscillator when taken low: when high or floating the astable runs normally.

The squarewave output from IC1c is differentiated by C26/R2 giving a negative pulse of a few tens of nanoseconds on IC1d output. This pulse can be used to slave several generators together for more complex applications. In this simple instrument, however, the pulse fires IC2 monostable, whose period is governed by VR2 and S2. The IC2 output is the *Pre-pulse*.

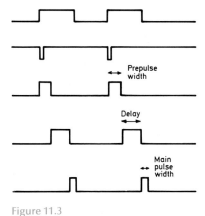

Figure 11.3

Waveforms

On completion of the pre-pulse, IC3 monostable fires to generate a time delay before firing IC4 at the end of the delay.

Fig. 11.3 shows the timing relative to IC1 clock. The pre-pulse is of width set by VR2/S2, the output pulse of width set by VR4/S4 is delayed from the pre-pulse by a delay set on VR3/S3. The overall operating frequency is governed by VR1/S1.

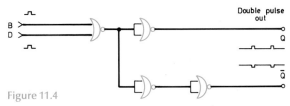

Figure 11.4

Double pulse out. All gates 74128

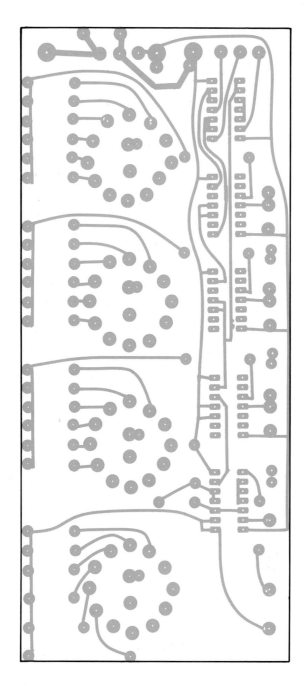

Figure 11.5
PCB layout. Drill four ⅜in holes on left to mount VR1/4

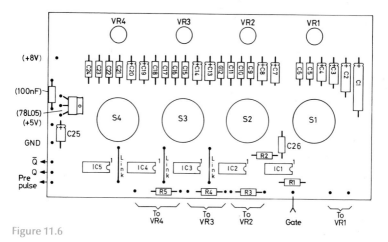

Figure 11.6

Component layout on PCB

In order to make the output pulses of universal application, 50Ω line drivers are used as the output devices rather than normal TTL. The 50Ω drivers are in IC5.

Construction

All components including the range switches and variable resistors can be mounted on the single PCB (Fig. 11.5) ensuring simple construction. The first stage is to solder the four

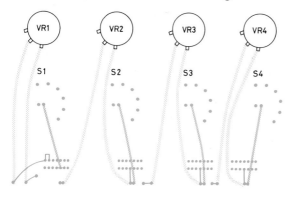

Figure 11.7

Connecting VR1/4 and S1/4 to the PCB. The lighter coloured lines connect VR1/4 to the PCB, while the darker coloured lines connect the S1/4 wipers to the PCB, where the circuit board is shown in solid colour

switches in position (Fig. 11.6) making sure that they are flat to the board surface. VR1 to VR4 can then be mounted in their respective holes on the PCB. Should 74128 be difficult to locate, 7402 could be used but the output capability will not be so good when driving co-ax test leads.

The board is then completed by inserting the i.c.s, resistors and capacitors, taking care to locate them correctly. Small solder pins are inserted for external connections to the board.

Twisted instrument wire is used to connect VR1 to the two pins adjacent to IC1. Similarly, VR2, VR3 and VR4 are connected to their respective points (Fig. 11.7). A finished view of the underside of the board is shown in Fig. 11.8.

Figure 11.8

Detail of underside of board, showing wiring to VR1/4 and S1/4

The wipers of the four range switches are connected to the PCB with a short length of instrument wire follows:

        S1 to IC1 pins 9, 10, 11
        S2 to IC2 pin 11
        S3 to IC3 pin 11
        S4 to IC4 pin 11

On board regulator

Power required for the pulse generator is +5V at around 100mA. This voltage must be accurately maintained and limited to an absolute maximum of +5.25V.

As an alternative arrangement, an on-board 78LO5 5V i.c. regulator can be inserted as shown on the layout, together with a 100nF capacitor. Power applied to the board can then be anything from 8V to 15V. At 10V or more a heatsink will be required for the regulator.

Double pulse

Some applications require the use of a double pulse pair of definable width and pulse separation. By gating the pre-pulse outputs in the circuit shown in Fig. 11.4, a double pulse output can be obtained. The i.c. is a 74128 line driver and the whole circuit can be built on a small piece of Veroboard adjacent to the double pulse output socket on the completed instrument. Power is applied ground to pin 7, +5V to pin 14, with a 100nF capacitor connected across the pins.

Mounting

Because of the single PCB design carrying all controls, housing the pulse generator is quite simple. Four ⅜in holes are drilled in the panel to support the threaded bosses of S1 to S4, which holds the PCB in place. Four ¼in holes are then drilled for the shafts of VR1 to VR4.

All that then remains is to mount the output switches on the panel together with a power switch and indicator.

Use

When 74128 output i.c.s have been used, the pulse generator will comfortably drive quite long co-ax lines between the generator and device on test.

N.B. 74128 devices can be quickly destroyed should the output be accidentally short circuited. 7402 devices are rather more robust and might be preferable for general purpose experimentation.

Setting up a pulse train is quite straightforward, particularly if an oscilloscope is used to verify the output.

(a) Set repetition frequency of pulse train
(b) Set width of pre-pulse – this cannot, of course, be longer than the period of repetition frequency.
(c) Set required delay.
(d) Set required output pulse width.

Table 11.1 Components for TTL pulse generator

I.C.s
IC1	7400
IC2	74121
IC3	74121
IC4	74121
IC5	74128

Resistors
R1	100Ω
R2	1kΩ
R3	1.5kΩ
R4	1.5kΩ
R5	1.5kΩ

(¼W unless stated)
VR1	1kΩ
VR2	25kΩ
VR3	25kΩ
VR4	25kΩ
PCB	

Capacitors
C1	470μF, 16V	C14	1μF, 16V
C2	47μF, 16V	C15	100nF
C3	4.7μF, 16V	C16	10nF
C4	0.47μF, 16V	C17	1nF
C5	47nF	C18	100pF
C6	4.7nF	C19	10μF, 16V
C7	10μF, 16V	C20	1μF, 16V
C8	1μF, 16V	C21	100nF
C9	100nF	C22	10nF
C10	10nF	C23	1nF
C11	1nF	C24	100pF
C12	100pF	C25	33μF, 6.3V
C13	10μF, 16V		

Miscellaneous
S1	One-pole, six-way *Lorlin*
S2	One-pole, six-way *Lorlin*
S3	One-pole, six-way *Lorlin*
S4	One-pole, six-way *Lorlin*

Driving other logic families

CMOS run on rails up to about 8V can be reliably driven from this generator. Other than that a simple level translator will be required to enable the high level output to exceed the switching threshold of the CMOS. Similarly ECL will require level translation. DTL logic will usually be TTL compatible and requires no modification.

Fig. 11.9 shows a trace of the pulse generator output at 50ns/cm.

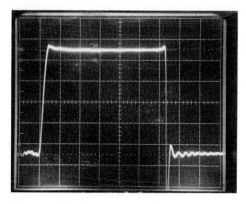

Figure 11.9

Pulse output at 50ns/cm

12

Total Harmonic Distortion Meter

Total harmonic distortion, or THD, is an important parameter relating to the linearity of any piece of amplification equipment. The meter described in this chapter will enable THD measurements to be carried out over the frequency range of 25Hz – 30kHz. Harmonic distortion down to 0.03% or less can be measured when used in conjunction with an audio oscillator of significantly better performance, such as the design presented in another chapter of this book.

As a readout the THD meter is used in conjunction with a millivoltmeter set to the 10mV range – visual examination of the THD residual waveform on a CRO is also beneficial.

Circuit

The output of an amplifier, when distorted, is made up of the desired output signal, say 1kHz, together with varying proportions of harmonics of that signal 2kHz, 3kHz 4kHz, etc. produced by the nonlinearity inherent in the amplifier design.

To assess the ratio of these unwanted harmonics, relative to the required output, as a percentage, the following method is employed.

A pure signal, containing minimal harmonics or noise, is fed to the unit under test (Fig. 12.1) and the distorted output fed to the THD meter. The signal level at the input to the THD meter is set to a reference level, representing a 100% signal, by the 'SET 100%' control. A notch filter is then tuned to reject completely the fundamental test signal, leaving as a residual the signal harmonics plus noise. The mean of this residual is then measured relative to the 100% signal and is expressed directly in percentage terms.

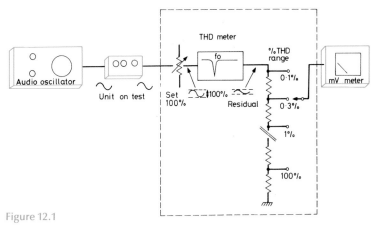

Figure 12.1
THD measurement

The basis of the instrument is the notch filter, which must provide a stable notch of very high rejection (>80dB) as well as being very selective, and attenuation should be zero at the second harmonic of the notch.

This design (Fig. 12.2) uses a 'twin tee' filter (Fig. 12.3) tuned in several ranges from 25Hz to 30kHz. Essentially a passive device, the twin tee is incorporated in an active configuration in this design in order to sharpen the notch and reduce attenuation at $2f_o$. The notch is fed from a fairly low impedance from VR1 etc. which sets the 100% level at around 300mV. IC1 provides a high impedance to the twin tee output and gives a gain of ×10. This gain is set by the potential divider R7/R8+R9.

A portion of the output of IC1, a little less than 1/10, or in fact a little less than the input to the stage, is fed to IC2 which provides a 'bootstrap' to the twin tee, sharpening the notch. S1 shorts out the twin tee so that the 100% reference level can be set.

The output at 100% from IC1 is about 3V, and this is attenuated by the chain connected to S2, the RANGE switch, to a level of 3mV when S2 is switched to the 100% position. IC3 raises this signal to a level of 10mV for the millivoltmeter to the output. When the notch is tuned the fundamental signal is removed. If the harmonics were 1%, then the level of the output from IC1 would be 1% of 3V: about 30mV. With S1 at the 1% position, the input to IC3 would be 3mV, the output to the millivoltmeter being 10mV or f.s.d.

Mains powered equipment, or test bench 'rat's nest' layouts, inevitably have a small degree of 50Hz and 100Hz hum in the

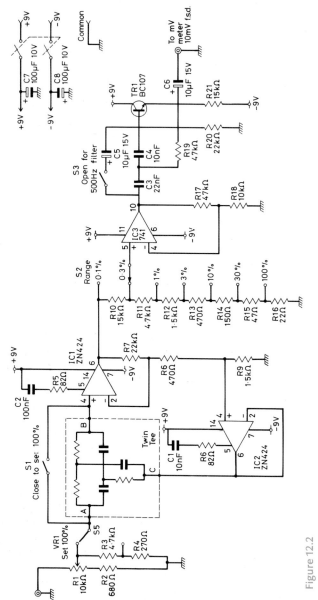

Figure 12.2
THD meter

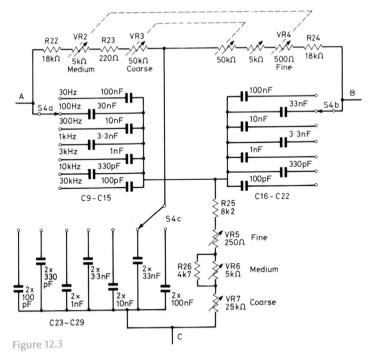

Figure 12.3

Twin tee

output, often more than the THD content. TR1 is connected as a simple 500Hz high pass filter which effectively removes such low frequencies from the output, giving a true reading of THD unaffected by hum in the signal.

Construction

The basic electronics is built on a 5in × 2½in piece of 0.1in *Veroboard*, strips running lengthwise.

The board (Fig. 12.4) is prepared by cutting the strips at the positions marked X on the layout with a special *Vero* cutter (total 23 cuts). Six wire links are inserted as shown.

The three i.c.s are soldered in place, taking care to insert them correctly, together with TR1 and the electrolytic capacitors, C6, C7 and C8. The remaining components are inserted and finally ten pins are soldered in place to take the external connections to the board.

At this stage the housing for the instrument should be considered. A metal enclosure should be chosen so that the instrument will be well screened – the instrument is basically a

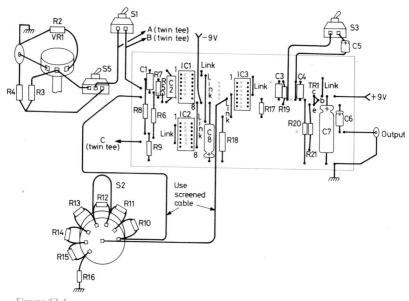

Figure 12.4

Veroboard layout. × = cut strip (total 23)

millivoltmeter, sensitive to 300µV f.s.d. There are nine rotary controls (Fig. 12.5) accommodated on the front panel in addition to three toggle switches and an ON/OFF switch, therefore a reasonably large enslosure is required.

Following the layout of Fig. 12.4, VR1, S1 and S5 are wired using screened cable if the wiring cannot be kept short and direct. R16 to R10 are mounted on the rear of S2 which is mounted on the front panel and wired to the *Veroboard*, again using screened cable.

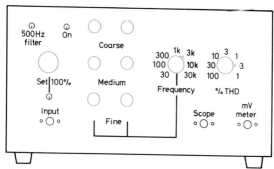

Figure 12.5

Suggested panel layout

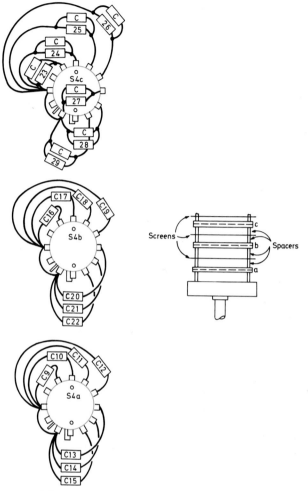

Figure 12.6
Switch S4

Switch S4

S4 is built up (Fig. 12.5) from three miniature switch wafers on a miniature switch mechanism. The twin tee capacitors are mounted on each wafer as shown in Fig. 12.6.

The wafers are shown with the twelve main contacts uppermost, with the two lower contacts to the left. Take particular care of this point – there should be *six* tabs to the right of the wafer and *eight* on the left hand side.

After each wafer has been assembled, the wafers are built up on the switch mechanism, separated by screening plates and spacers. Adjust the end stops on the switch mechanism to allow the wafer to cover the correct seven locations.

Twin tee controls

S4 is wired to the control pots for the twin tee as in Fig. 12.7. Short direct wiring should be used to connect the pots and S4 to the *Veroboard*: the three twin tee connections are identified as A, B and C on the drawings. A and B are interchangeable. Ideally the pots are spaced on the front panel so that the resistors R24, R22, R23 and R25 connect directly without the need for extending the leads.

Finally, the batteries are connected via a double pole ON/OFF switch and the output socket connected (In point of fact it is useful to provide a duplicate output so that an oscilloscope can be connected to examine the residual).

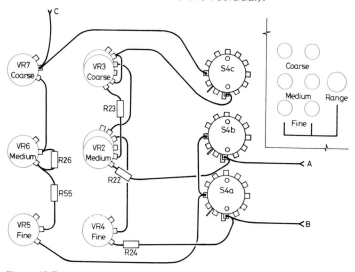

Figure 12.7

Layout of twin tee

Use

It is essential that the test signal THD is below that expected from the equipment on test by a factor of at least two, preferably more. At 1kHz the audio oscillator design in this book would allow accurate THD measurements at 0.01% or more.

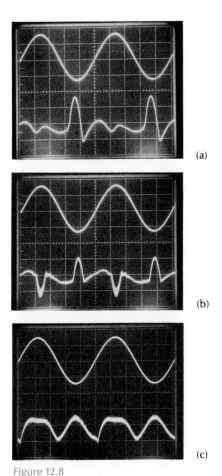

Figure 12.8

THD residuals. (a) Amplifier clipping – asymmetrical, negative half cycle clipping. (b) Amplifier clipping – positive and negative half-cycles. (c) Crossover distortion

The test signal is applied to the circuit being tested and the levels etc. adjusted so that the measurement conditions are met (e.g. 1kHz, 10W at 8Ω). The THD meter is then connected across the output and the SET 100% control adjusted so that, with the SET 100% switch closed and the range set at 100%, the millivoltmeter reads full scale. The SET 100% switch is then opened, passing the signal through the twin tee. The FREQUENCY switch S4 is then set to the appropriate range and in pairs the COARSE, MEDIUM and FINE controls are tuned to give minimum deflection on the meter. The RANGE is then

Table 12.1 Components for total harmonic distortion meter

Resistors

R1	10kΩ	R14	150Ω	2%
R2	680Ω	R15	47Ω	2%
R3	4.7kΩ	R16	22Ω	2%
R4	270Ω	R17	47kΩ	
R5	82Ω	R18	10kΩ	
R6	82Ω	R19	47kΩ	
R7	22kΩ	R20	22kΩ	
R8	470Ω	R21	15kΩ	
R9	1.5kΩ	R22	18kΩ	
R10	15kΩ 2%	R23	220Ω	
R11	4.7kΩ 2%	R24	18kΩ	
R12	1.5kΩ 2%	R25	8.2kΩ	
R13	470Ω 2%	R26	4.7kΩ	

I.C.s

IC1	ZN424
IC2	ZN424
IC3	741

Transistor

TR1	BC107

Capacitors

C1	10nF	C16	100nF
C2	10nF	C17	33nF
C3	22nF	C18	10nF
C4	10nF	C19	3.3nF
C5	10μF, 15V	C20	1nF
C6	10μF, 15V	C21	330pF
C7	100μF, 10V	C22	100pF
C8	100μF, 10V	C23	2 × 100nF (in parallel)
C9	100nF	C24	2 × 33nF
C10	33nF	C25	2 × 10nF
C11	10nF	C26	2 × 3.3nF
C12	3.3nF	C27	2 × 1nF
C13	1nF	C28	2 × 330pF
C14	330pF	C29	2 × 100pF
C15	100pF		

Switches

S1	Single pole toggle
S2	1P 7W rotary
S3	Single pole toggle
S4	3P 7W, built from miniature switch kit, RS Components, Maplin etc.
S5	Double pole toggle

Controls
VR1	10kΩ lin. pot,
VR2	5kΩ lin. pot, 2 Gang
VR3	50kΩ lin. pot, 2 Gang
VR4	500Ω lin. pot,
VR5	250Ω lin. pot,
VR6	5kΩ lin. pot,
VR7	25kΩ lin. pot,

Miscellaneous
Veroboard 5in × 2½in 0.1in pitch
Suitable case (metal for screening)
Input and output sockets (BNC)
2 × PP9 batteries
Battery ON/OFF switch. (DP)

reduced and the tuning process continued to obtain the very minimum reading. By noting the reading on the RANGE switch the meter then reads THD in percent.

At frequencies in excess of 1kHz, S3 can be opened, switching in the 500Hz filter – reducing the effect of hum on the reading.

Additional to the millivoltmeter, an oscilloscope connected to the output gives very useful visual clues as the nature of the distortion, as well as making the tuning so much easier.

Extensions

It is quite a simple matter to make the THD meter self-contained by incorporating the millivoltmeter of Chapter 2 into the same case as the THD electronics. The millivoltmeter can either be modified to a have a single sensitivity of 10mV, or alternatively the complete unit can be built in, with connections on the front panel to connect to the THD meter or use as a millivoltmeter.

Mains power for the THD meter is not really a good idea – problems arise from earth loops introducing mains hum into the measurements.

The sensitivity of the instrument is such that input signals as low as 300mV can be used for measurements. Should a lower signal be available, it is quite simple to set 100% at, say, the 10% setting of the RANGE switch. The most sensitive THD range is then 1% f.s.d. Alternatively a pre-amplifier can be switched into circuit, but it is clearly necessary that the THD and noise of the pre-amplifier is sufficiently low as not to give misleading measurements.

Appendix

Making PCBs

Printed circuit boards, as used for many of the projects in this book, are made up of a thin base of fibreglass or S.R.B.P. (synthetic resin bonded paper) with a layer of copper coated on one side. The copper is etched to leave a conductor pattern which connects various components which are inserted into holes in the PCB and soldered to the copper tracks. PCBs have the big advantage of ensuring repeatable results and neat layouts. Construction time is reduced to the time necessary to insert and solder the components.

Step-by-step guide

1. Cut a piece of laminate to the size of the final PCB with a small hacksaw. File the edges smooth and polish the copper surface clean with a special PCB cleaning block or fine wire wool. It is essential to remove all grease stains, fingerprints and tarnish.

2. Examine the board layout and identify any components of critical location or pin spacing, such as mounted switches, i.c.s, etc. From layout provided, accurately trace the location of these components onto tracing paper. Position this tracing accurately over the prepared copper surface of the board and transfer the component location to the copper by carefully pricking through the tracing with a schoodboy's compass or scribe. (For i.c. pins it is better just to locate pin 1 in this way and then use a ruler to space the remaining pins accurately at 0.1in intervals, separated by 0.3in spaces.)

3. Using a Decon Laboratories *Dalo* pen, very carefully copy the whole of the required track layout onto the copper (Fig. A.1). The Dalo pen is rather like a felt tip pen with a valve and releases a thick etch-resistant ink. Take special care with adjacent pads, such as i.c. pins, which are so close that the inks may run over to an adjacent pad. Any such errors can be rectified after the ink has dried by chipping away with a small sharp knife.

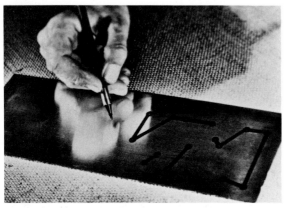

Figure A.1

How the Decon Laboratories 'decon-dalo' PCB pen is used to draw the track layout onto the copper side of the board

4. When the resistant ink is fully set (about 4 hours minimum), the board can be etched.

> *CAUTION: Ferric chloride is a powerful etchant for most materials except glass and some plastics.*
> *TAKE CARE — MOP UP ALL SPILLS PROMPTLY AND WEAR HOUSEHOLD RUBBER GLOVES. AVOID SPLASHING CLOTHES.*

An etching solution of ferric chloride is made up from ferric chloride crystals dissolved in water to make a saturated solution. If 500g of crystals are used, filling the glass container (in which they are supplied) with hot water will make up just such a saturated solution. Alternatively ready prepared etchant fluid can be purchased at a higher cost.

A one inch depth of etchant fluid in a photographic developing dish is sufficient to etch several PCBs. The board is placed in the liquid face up and the dish continuously agitated. Eventually, the copper nearest to the edge of the board and also nearest to large areas of resist will etch through to the fibreglass. Progressively all exposed copper will be removed. At this stage the board is removed from the solution and washed. The solution is then stored for further use until it is a dense black colour, when developing will take so long as to be quite useless, and a new solution will be required.

5. The etch-resist ink is removed from the etched board either by washing off with acetone, or scrubbing with fine wire wool. All that remains now is to drill the board. Most of the board will need to be drilled 1mm (or no. 60 drill) but large value capacitors, preset resistors and certain other components need larger holes. At this stage it is as well to drill any mounting holes that may be required.

Finally, hold the board against a bright light and check that none of the tracks have breaks in them. Small gaps can be bridged with a blob of solder quickly applied, while 22 s.w.g. wire will bridge large breaks.

Take care to keep the board dry before construction, otherwise the copper will tarnish and soldering will be difficult. If boards are to be stored for a while it is a wise precaution to spray the copper side lightly with PCB preservative first.

Newnes
constructors projects

☆ More books in our series for the electronics enthusiast
☆ There are now eight titles available

    Electronic Projects in the Workshop *R.A. Penfold*
    Electronic Projects in Music *A.J. Flind*
    Electronic Projects in Audio *R. A. Penfold*
    Electronic Projects in Hobbies *F.G. Rayer*
    Electronic Projects in the Car *M. George*
    Electronic Game Projects *F.G. Rayer*
    Electronic Projects in the Home *Owen Bishop*
    Projects in Radio and Electronics *Ian R. Sinclair*

Each book contains a collection of constructional projects, giving details of how the circuit works, how it may be assembled and how setting-up and trouble-shooting problems may be solved. The skilful use of colour in the text helps to clarify circuit operation, and circuit board layouts are suggested. Shopping lists of components are included.

Write for a free colour brochure about all our hobby books to

Newnes Technical Books
Borough Green, Sevenoaks, Kent TN15 8PH

Practical Electronics Handbook

Ian Sinclair

A useful and carefully selected collection of standard circuits, rules-of-thumb and design data.

Covers passive and active components, discrete component circuits and linear and digital i.c.s.

Describes the operation and function of typical circuits whilst keeping mathematics to a minimum.

208 pages   216 x 138mm   0 408 00447 9

Newnes Technical Books
Borough Green, Sevenoaks, Kent TN15 8PH

Beginner's Guide to Digital Electronics
Ian Sinclair

Digital electronics affects us all – pocket calculators, digital watches, T.V. games, microprocessors and computers all make use of digital technology. This book provides a readable introduction to the methods, circuits and applications of digital electronics, with practical hints and exercises.

1980   192 pages   0 408 00449 5

Electronics – Build & Learn
R A Penfold

An introduction to basic electronics theory for beginners by means of practical experiments: 'learning by doing'. Full instructions are given for building a circuit demonstrator unit; electronics components are then described one by one and used in simple circuits that can be set up on the unit. This is an ideal first book for hobbyists.

1980   112 pages   0 408 00454 1

Newnes Technical Books
Borough Green, Sevenoaks, Kent TN15 8PH